- **The larger the set of consecutive prime numbers, the narrower the search for large prime numbers**

My discovered formula:

Definitions:

Array PTBP

It is the following Array of odd numbers

$$\begin{vmatrix} 1 & 3 & 7 & 9 \\ 11 & 13 & 17 & 19 \\ 21 & 23 & 27 & 29 \\ 31 & 33 & 37 & 39 \\ 41 & 43 & 47 & 49 \\ 51 & 53 & 57 & 59 \end{vmatrix}$$

And so on....

Name: Eng. / Mostafa Yacoub
Abdellatif Mahmoud

Nationality: Egyptian

ORCID: 0000-0002-9991-4624

Email:

moshhaabma2015@gmail.com

Qualification: civil engineer
Cairo University 2003

- **<u>Encircle prime numbers</u>**
- In this paper or research, we will explain the distribution of prime numbers that have the last digit = 1 , 3   7, or 9 based on my discovered formula that connects prime and composite numbers.
- We will narrow the scope of the search for prime numbers according to the recurring patterns resulting from a specific set of consecutive prime numbers and the composite numbers that result from them

- **For a given set of consecutive primes whose numbers =n that start with prime 3 and end with prime F and not including prime 2 and prime 5**
  i.e.
  set=[3,7,11,13,………………………………
  ………………………………,F]
  S=product of those consecutive primes

i.e

$$S = \prod_{i=3}^{i=F} (i)$$

Range=R_k = 10 × S × k

Where k = [1, 2, 3, 4, ..., ∞(infinity)

i.e R_1=10 x S x 1 and R_2=10 x S x 2

And so on

- **Number of composite numbers that belong to Array PTBP and created by the effect of those consecutive primes within the range R_K**

- $$=\left[\left(K \times 4^{\times \frac{S}{3}}\right)+\left(\sum_{j=7}^{j=F}\left(K \times 4 \times \left(\frac{S}{j}\right) \times \prod_{i=7}^{i=\text{prime number befor current prime number } j}\left(\frac{i-1}{i}\right)\right)\right)\right]-(n)$$

Where j =consecutive values of primes

7, 11, 13,..............., F

And i= consecutive values of primes

3, 7, 11, 13,........, prime before current j prime

- **The previous formula can be applied for any number of consecutive prime numbers that start with prime number 3**

- **The first term** $(k \times 4 \times \frac{S}{3})$ represents the count of unique Composite numbers +1 that belong to the Array PTBP and are created by prime number 3 within the range

$$R_k = 10 \times S \times k$$

- **The second term**

$$\sum_{j=7}^{j=F} (K \times 4 \times (\frac{S}{j}) \times$$

$i = prime\ number\ befor\ current\ prime\ number\ j$

$$\prod_{i=7} \qquad (\frac{i-1}{i})$$

Represent the count of unique Composite numbers+n-1 that belong to the Array PTBP and are created by each prime number after the prime number 3 within the range

$R_k = 10 \times S \times k$

- **The first term** $(k \times 4 \times \frac{S}{3})$
  represents the count of unique
  Composite numbers +1 that belong
  to the Array PTBP and are created
  by prime number 3 within the
  range

$$R_k = 10 \times S \times k$$

- **The second term**

$$\sum_{j=7}^{j=F} (K \times 4 \times (\frac{S}{j}) \times$$

$i = prime\ number\ befor\ current\ prime\ number\ j$

$$\prod_{i=7} \qquad (\frac{i-1}{i})$$

Represent the count of unique Composite numbers+n-1 that belong to the Array PTBP and are created by each prime number after the prime number 3 within the range

$R_k = 10 \times S \times k$

- **The third term (-n)**

  Subtracting n (number of consecutive primes starting from prime number 3) because the count of composite numbers generated from those consecutive primes includes the count of those primes in the range

  $R_k = 10 \times S \times k$

- Explanation and proof for my theory in my previous paper (prime number theory)
- We will mention only the concept of number cycle
We can use the number cycle concept to understand the behavior of consecutive primes in creating composite numbers.

i.e.

$$S = \prod_{i=3}^{i=F} (i)$$

Range=cycle range= R_k = 10 × S × k

Where k= [1, 2, 3, 4, ...,

∞(infinity)

i.e. R_1=10 x S x 1 and R_2=10 x S x 2

And so on

- **Now consider only one k value =1**

For a set of consecutive primes and according to my formula the result will be

- $$= \left[\left(K \times 4^{\times \frac{S}{3}} \right) + \left(\sum_{j=7}^{j=F} \left(K \times 4 \times \left(\frac{S}{j} \right) \times \prod_{i=7}^{i=\text{prime number befor current prime number } j} \left(\frac{i-1}{i} \right) \right) \right] - (n)$$

And

including the count of prime numbers within the set

$$=[(K \times 4^{\times \frac{S}{3}}) + (\sum_{j=7}^{j=F} (K \times 4 \times (\frac{S}{j}) \times$$

$i = $ *prime number befor current prime number j*

$$\prod_{i=7} (\frac{i-1}{i})$$

)]

- **Which represent the count of numbers (that belong to the Array PTBP) that is divisible of the prime numbers that belong to the set of consecutive primes**

$t$he complementary part

$$= (4 \times \prod_{i=3}^{i=F} (i-1))$$

As explained in my paper (infinite primes)

Which represent the count of numbers (that belong to the Array PTBP) that is not divisible by each prime number that belong to the set of consecutive primes

- For example if the set of consecutive primes=[3] then we have the following pattern of uncolored numbers (except 1) that continues to infinity
- and we can Define and limit the scope of the search for prime numbers within that repeated pattern

1	3	7	9
11	13	17	19
21	23	27	29
31	33	27	39
41	43	27	49
51	53	27	59
61	63	27	69
71	73	27	79
81	83	27	89
91	93	27	99
101	103	27	109
111	113	27	119

- And this pattern continues to infinity
- And now if the set of consecutive primes=[3,7] then we have the following more accurate repeated pattern of uncolored numbers (except 1) that continues to infinity
- Which limits the process of searching for prime numbers to more precise places
- than the previous set that has only the prime number 3

1	3	7	9
11	13	17	19
21	23	27	29
31	33	37	39
41	43	47	49
51	53	57	59
61	63	67	69
71	73	77	79
81	83	87	89
91	93	97	99
101	103	107	109
111	113	117	119

121	123	127	129
131	133	137	139
141	143	147	149
151	153	157	159
161	163	167	169
171	173	177	179
181	183	187	189
191	193	197	199
201	203	207	209
211	213	217	219
221	223	227	229
231	233	237	239
241	243	247	249
251	253	257	259
261	263	267	269

271	273	277	279
281	283	**287**	289
291	293	297	299
301	303	307	309
311	313	317	319
321	323	327	**329**
331	333	337	339
341	**343**	347	349
351	353	357	359
361	363	367	369
371	373	377	379
381	383	387	389
391	393	397	399
401	403	407	409
411	**413**	417	419

421	423	**427**	429
431	433	437	439
441	443	447	449
451	453	457	459
461	463	467	**469**
471	473	477	479
481	483	487	489
491	493	**497**	499
501	503	507	509
511	513	517	519
521	523	527	529
531	533	537	**539**
541	543	547	549
551	**553**	557	559
561	563	567	569

571	573	577	579
581	583	587	589
591	593	597	599
601	603	607	609
611	613	617	619
621	623	627	629

- And this pattern continues to infinity
- As we add more consecutive prime numbers to the set we get a more accurate pattern of uncolored numbers that continues to infinity Which limits the process of searching for prime numbers to more precise places

- and the percentage of those uncolored numbers decreases

relative to the range of that set for all values of k from 1 to infinity

- **And we can define that percentage to be equal to**

$$[(4 \times \prod_{i=3}^{i=F} (i-1))] / [4 \times \prod_{i=3}^{i=F} (i)]$$

$$=[(\prod_{i=3}^{i=F} (i-1))] / [\prod_{i=3}^{i=F} (i)]$$

$$\prod_{i=3}^{i=F} [(i-1)/i]$$

Where F is the last prime number in the set